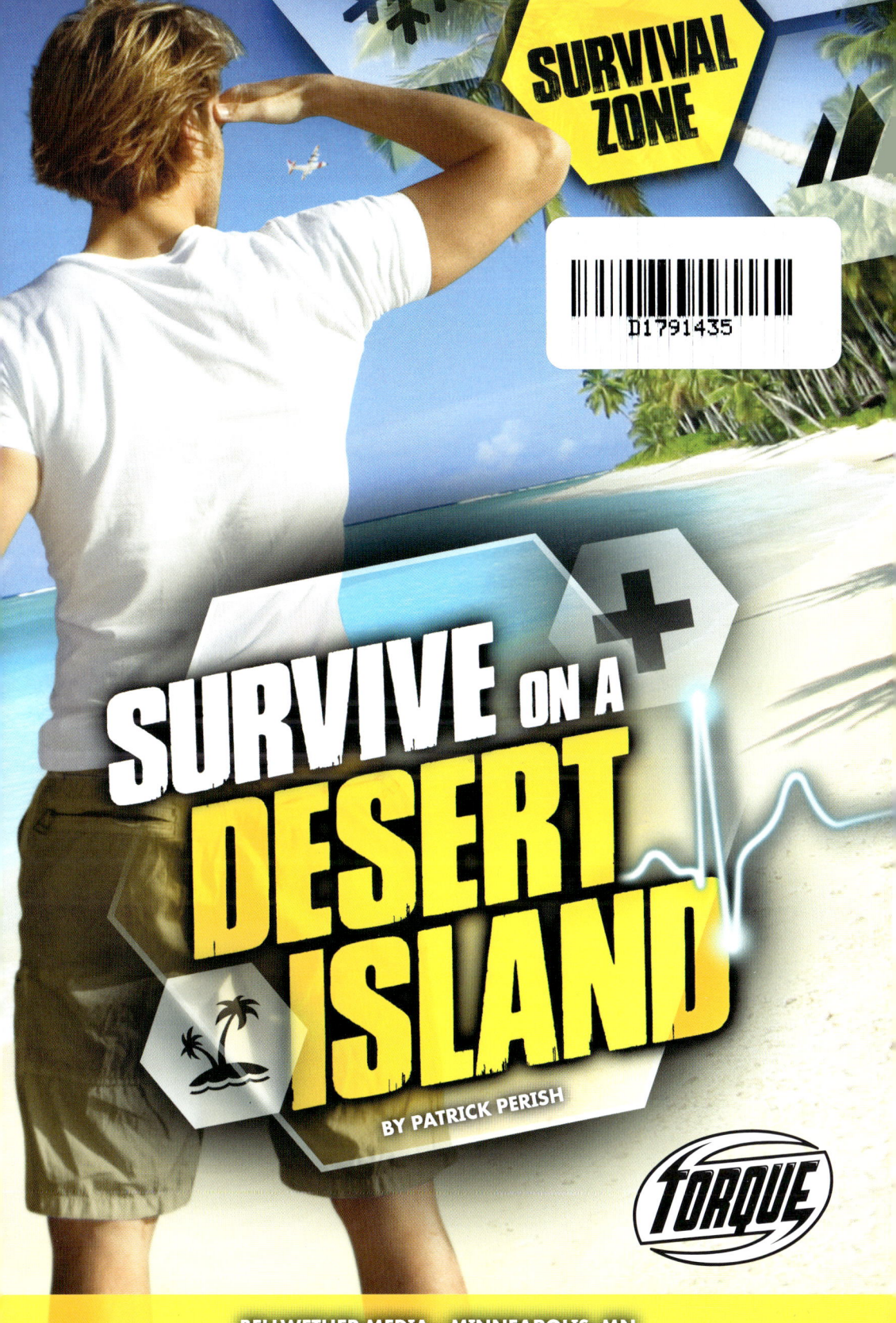

SURVIVAL ZONE

SURVIVE ON A DESERT ISLAND

BY PATRICK PERISH

TORQUE

BELLWETHER MEDIA · MINNEAPOLIS, MN

Are you ready to take it to the extreme? Torque books thrust you into the action-packed world of sports, vehicles, mystery, and adventure. These books may include dirt, smoke, fire, and chilling tales. **WARNING**: read at your own risk.

This edition first published in 2017 by Bellwether Media, Inc.

No part of this publication may be reproduced in whole or in part without written permission of the publisher. For information regarding permission, write to Bellwether Media, Inc., Attention: Permissions Department, 5357 Penn Avenue South, Minneapolis, MN 55419.

Library of Congress Cataloging-in-Publication Data

Names: Perish, Patrick, author.
Title: Survive on a Desert Island / by Patrick Perish.
Description: Minneapolis, MN : Bellwether Media, Inc., 2017. | Series: Torque: Survival Zone | Includes bibliographical references and index.
Identifiers: LCCN 2016000247 | ISBN 9781626174481 (hardcover : alk. paper)
Subjects: LCSH: Desert survival–Juvenile literature.
Classification: LCC GV200.5 .P367 2017 | DDC 613.6/9–dc23
LC record available at http://lccn.loc.gov/2016000247

Text copyright © 2017 by Bellwether Media, Inc. TORQUE and associated logos are trademarks and/or registered trademarks of Bellwether Media, Inc.

SCHOLASTIC, CHILDREN'S PRESS, and associated logos are trademarks and/or registered trademarks of Scholastic Inc.

Printed in the United States of America, North Mankato, MN.

TABLE OF CONTENTS

Capsized! 4

Alone in the Wilderness 8

Water and Shelter 12

Food and Fire 16

Rescue! 20

Glossary 22

To Learn More 23

Index 24

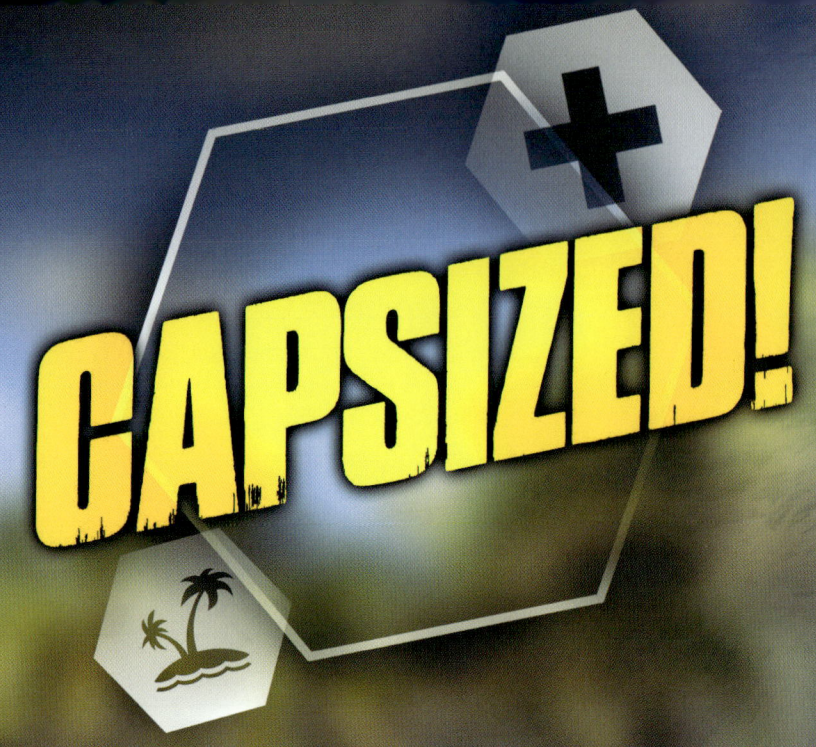

CAPSIZED!

On July 6, 2004, 11-year-old Stephen Nona and his family were headed to a birthday party. They left their home island near northern Australia and sailed for Thursday Island.

Partway into the journey, the engine died. Rough waves slammed into their boat, causing it to **capsize**! Stephen's parents told their kids to swim for rocks in the distance.

"[Stephen] said to the girls, 'If we don't swim for that island, we're going to die.'"
-Vickie Tamwoy, aunt of Stephen, Ellis, and Norita

"We swam all day. We started in the morning and we got to the big island in the afternoon."
-Stephen Nona

Stephen and his sisters, Ellis and Norita, swam for hours through shark-filled waters. Eventually, they reached the rocks.

The land offered little to eat or drink. The kids were forced to swim to a bigger island. There they found coconuts, wild fruit, and oysters. After being **stranded** for six days, they were rescued by their uncle. They had survived on a desert island!

ALONE IN THE WILDERNESS

A desert island is an island without people living on it. This might sound like a dream vacation. But it can quickly become a nightmare. Bad weather, little water, and even minor injuries can turn deadly.

Years ago, sailors were **marooned** on desert islands as punishment. Today, survivors of plane crashes or shipwrecks can find themselves stranded.

NO CACTI HERE

Desert islands are not deserts. In fact, many are in rainy areas.

Desert island survival depends on staying positive and getting creative. **Salvage** what you can from the **wreckage**. Collect **debris** that washes ashore.

One salvaged piece might have many uses. A rubber raft could provide shelter or catch rain. A piece of sharp metal might work as a knife.

REAL-LIFE CASTAWAY

In the 1700s, sailor Alexander Selkirk was stranded on a desert island. The famous book Robinson Crusoe was based on his story.

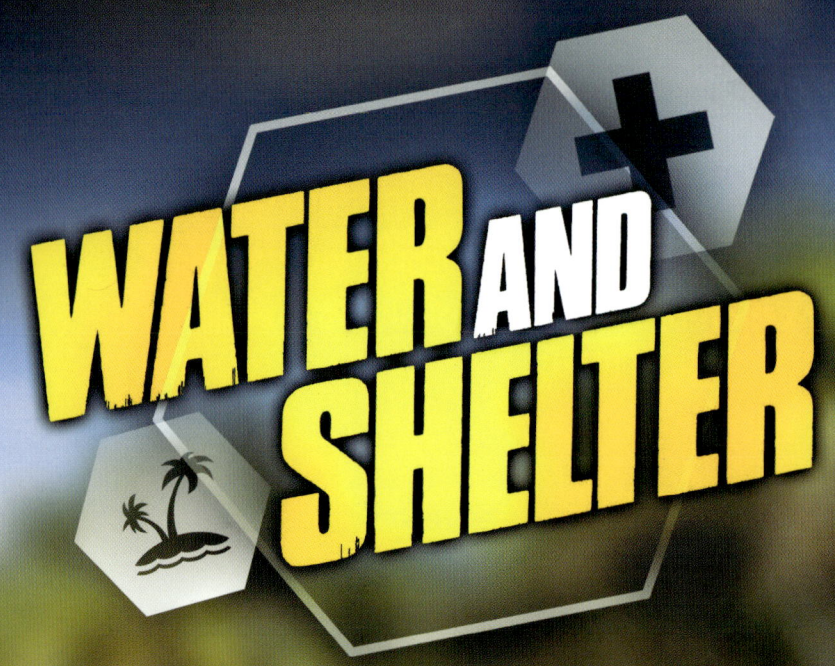

WATER AND SHELTER

It is important to find freshwater. During hot weather or heavy activity, a person can quickly become **dehydrated**. Seawater is too salty. Drinking too much will kill you. Search the island for streams. Find something to catch rain, such as boots or large leaves. Tie a plastic bag around leafy branches to collect water from plants.

THE RULE OF THREES

The rule of threes is a helpful reminder of basic needs. Humans can survive three minutes without air, three days without water, and three weeks without food.

Desert islands often receive a lot of rain. Without shelter to keep you warm and dry, you can develop **hypothermia**. Build a shelter near your water source. Keep it in sight of the ocean so rescuers can spot you.

Bamboo makes a strong frame. Layer palm leaves for a waterproof roof. Inside, create a raised bed to stay dry and away from snakes.

TYPES OF SHELTERS

lean-to

A-frame

tepee

debris hut

HEADS UP!
Avoid building your shelter too close to palm trees. A falling coconut can be deadly.

FOOD AND FIRE

Many islands have plenty of food. Eat only familiar fruits like bananas and coconuts. Strange berries may be poisonous. Look for fish in **tide pools**. Use a homemade fishing rod or sharpened stick to catch them. **Shellfish** can be gathered from the shallows. If you have a fire, you should cook any insects or other animals you eat.

ISLAND FOOD

 fish

 mussels

crayfish

 seaweed

 bananas

 coconuts

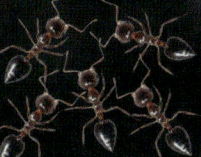 ants

 grubs

BUG APPETIT!

Bugs are a great source of energy. Many people around the world eat insects every day.

A fire can make island life much easier. It can cook food and help rescuers find you. First, gather dry **tinder** and **kindling**.

If you could not salvage a lighter, there are still ways to make flames. You can make a bow drill from dry sticks and wood. A lens or thick glasses can focus sunlight into a hot beam. Scraping steel on **flint** can create a spark.

MAKING A BOW DRILL

A bow drill can start a fire if you do not have matches or a lighter.

MATERIALS: two flat pieces of dry wood, a sturdy stick, a bendy stick (bow), string, tinder

1. Carve a circle into each flat board to fit the ends of the stick
2. Attach the string to each end of the bow
3. Loop the string around the stick
4. Hold the larger flat piece on the ground with your foot
5. Fit the stick into the circle, and keep it upright with the other flat wood piece
6. Spin the stick by moving the bow back and forth until it starts smoking
7. Place tinder near the smoke to help the fire catch

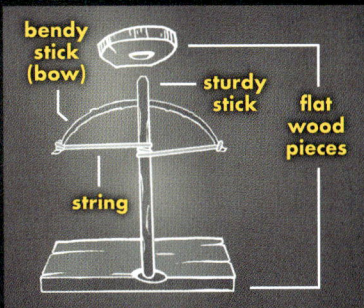

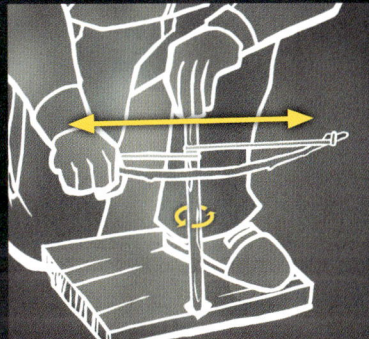

A CHILLY STAY

Canada's Devon Island is the largest desert island on Earth. You do not want to be stranded there. It is near the North Pole.

RESCUE!

Even with food, water, and shelter, desert islands are dangerous. It is important to prepare several rescue signals. Always be ready to signal at passing ships or planes.

Fires from green plants make smoky signals. Mirrors or shiny metal can reflect sunlight at planes or boats. Stay positive and be patient. With luck, you can survive on a desert island!

MAYDAY!

These rescue signals are recognized all over the world.

three rocks or fires in a triangle formation

SOS written out on the ground

large V made with rocks or branches

GLOSSARY

capsize—to turn over in the water

debris—the remains of something broken down or destroyed

dehydrated—a condition in which the body has lost too much water or fluid

flint—a hard gray rock that makes sparks when struck by steel

hypothermia—a condition in which the body loses heat faster than it can produce it; hypothermia causes body systems to shut down.

kindling—small sticks and twigs used to fuel a fire

marooned—to be left or trapped somewhere that is hard to get away from, usually an island

salvage—to save or gather, usually from a wreck

shellfish—animals with hard outer shells that live in water

stranded—left in a place with no way to get out

tide pools—rocky pools of water left after ocean waves retreat

tinder—dry grass, bark, or leaves used to start a fire

wreckage—the remains of something that has been damaged or destroyed

TO LEARN MORE

AT THE LIBRARY

Bell, Samantha. *How to Survive on a Deserted Island.* Mankato, Minn.: Child's World, 2015.

O'Dell, Scott. *Island of the Blue Dolphins.* Boston, Mass.: Sandpiper, 2010.

Pipe, Jim. *How to Survive on a Desert Island.* New York, N.Y.: PowerKids Press, 2013.

ON THE WEB

Learning more about surviving on a desert island is as easy as 1, 2, 3.

1. Go to www.factsurfer.com.

2. Enter "survive on a desert island" into the search box.

3. Click the "Surf" button and you will see a list of related web sites.

With factsurfer.com, finding more information is just a click away.

INDEX

Australia, 4
boat, 4, 20
bow drill, 18, 19
dangers, 8, 12, 14, 15, 16
debris, 10, 11
Devon Island, Canada, 19
fire, 16, 18, 19, 20, 21
food, 7, 13, 16, 17, 18, 20
hypothermia, 14
Nona, Ellis, 5, 6
Nona, Norita, 5, 6
Nona, Stephen, 4, 5, 6
plane, 8, 20
quotes, 5, 6
rain, 9, 11, 12, 14
rescue, 7, 14, 18, 20, 21
Robinson Crusoe, 11
safety, 12, 14, 15, 16
Selkirk, Alexander, 11
shelter, 11, 14, 15, 20
signals, 20, 21
tide pools, 16
tools, 11, 12, 16, 18, 19, 20
water, 6, 8, 12, 13, 14, 20
weather, 8, 9, 12, 14

The images in this book are reproduced through the courtesy of: Tyler Durden / Corbis Images, front cover (person); photogerson, front cover (background); Jodie Alaine, pp. 4-5 (waves); Giorgio Fochesato, p. 5 (boat); altanaka, pp. 6-7; Wouter Tolenaars, pp. 8-9; Design Pics Inc/Alamy, pp. 10-11; Chris Hellier/ Alamy, p. 11; Comaniciu Dan, pp. 12-13; Phoenix490, pp. 14-15; Visualisty, p. 15 (top left); Simon Gurney, p. 15 (top right); Kira Volkov, p. 15 (bottom left); Bruce Raynor, p. 15 (bottom right); Miss Aumada Yuenwong, p. 16; Dan loei tam loei, pp. 16-17; valzan, p. 17 (fish); Boltenkoff, p. 17 (mussels); Vadym Zaitsev, p. 17 (crayfish); Jiang Hongyan, p. 17 (seaweed); Maks Narodenko, p. 17 (bananas); MaraZe, p. 17 (coconuts); schankz, p. 17 (ants); Kovalchuk Oleksandr, p. 17 (grubs); ChameleonsEye, p. 18; Photo Image, pp. 18-19; Dave Wheeler, p. 19 (top, bottom); BlurOrange Studio, p. 20; Olleg, pp. 20-21; anathomy, p. 21 (beach); EkDanilishina, p. 21 (fires); I am Gavrila, p. 21 (rock V); Ivan Pavlov, p. 21 (SOS); kavalenkau, p. 21 (boat).